A.P. 2062A—P.N. (3rd Edition)
A.P. 2062C—P.N. (2nd ")
A.P. 2062F—P.N. (1st ")

ROYAL AIR FORCE
PILOT'S NOTES
for
LANCASTER
I, III & X

MARK I—FOUR MERLIN XX, 22 or 24 ENGINES
MARK III & X—FOUR MERLIN 28 or 38 ENGINES

By Royal Air Force

PROMULGATED BY ORDER OF THE AIR COUNCIL

©2013 Periscope Film LLC
All Rights Reserved
ISBN#978-1-937684-52-5
www.PeriscopeFilm.com

DISCLAIMER:

This manual is sold for historic research purposes only, as an entertainment. It contains obsolete information and is not intended to be used as part of an actual operation or maintenance training program. No book can substitute for proper training by an authorized instructor.

©2013 Periscope Film LLC
All Rights Reserved
ISBN#978-1-937684-52-5
www.PeriscopeFilm.com

AMENDMENTS

Amendment lists will be issued as necessary and will be gummed for affixing to the inside back cover of these notes.

Each amendment list will include all current amendments and will, where applicable, be accompanied by gummed slips for sticking in the appropriate places in the text.

Incorporation of an amendment list must be certified by inserting date of incorporation and initials below

A.L. No.	INITIALS	DATE	A.L. No.	INITIALS	DATE
1			7		
2			8		
3			9		
4			10		
5			11		
6			12		

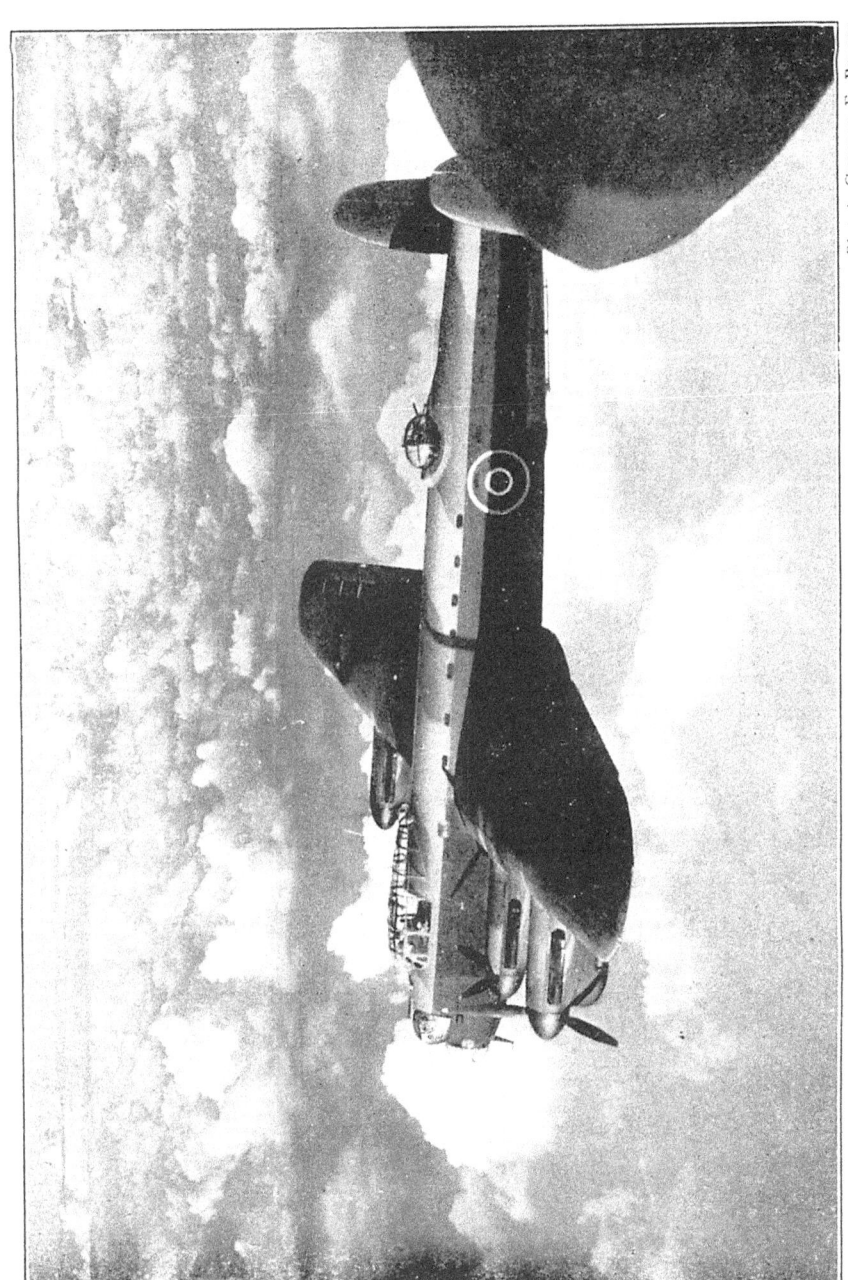

Photo by Charles E. Brown

LANCASTER I, III & X

NOTES TO USERS

This publication is divided into six parts: Descriptive, Handling, Operating Data, Emergencies, Supplementary Notes for Flight Engineer, and Illustrations and Location of Controls.

These Notes are complementary to A.P. 2095 Pilot's Notes General and assume a thorough knowledge of its contents. All pilots should be in possession of a copy of A.P. 2095 (see A.M.O. A93/43). Flight Engineers should also have a copy of A.P. 2764 when issued.

Words in capital letters indicate the actual markings on the controls concerned.

Additional copies may be obtained from A.P.F.S., Fulham Road, S.W.3, by application on R.A.F. Form 294A, in duplicate, quoting the number of this publication in full— A.P. 2062A, c and f—P.N.

Comments and suggestions should be forwarded through the usual channels to the Air Ministry (D.T.F.).

AIR MINISTRY AIR PUBLICATION 2062A, C, & F—P.N.
May 1944 *Pilot's and Flight Engineer's Notes*

LANCASTER I, III & X
PILOT'S & FLIGHT ENGINEER'S NOTES

LIST OF CONTENTS

PART I—DESCRIPTIVE *Para.*

INTRODUCTION .. 1

FUEL AND OIL SYSTEMS
- Fuel tanks .. 2
- Fuel cocks .. 3
- Vapour vent system (Lancaster III and X aircraft only) .. 4
- Electric fuel booster pumps .. 5
- Fuel contents gauges .. 6
- Fuel pressure indicators .. 7
- Priming pumps .. 8
- Oil tanks .. 9
- Oil dilution .. 10

MAIN SERVICES
- Hydraulic system .. 11
- Pneumatic system .. 12
- Electrical system .. 13

AIRCRAFT CONTROLS
- Trimming tabs .. 14
- Undercarriage control .. 15
- Undercarriage indicator .. 16
- Undercarriage warning horn .. 17
- Flaps control .. 18
- Bomb doors .. 19

ENGINE CONTROLS
- Throttle controls .. 20
- Mixture control .. 21
- Propeller controls .. 22

PART I—*continued* *Para.*

    Supercharger controls 23
    Radiator shutters 24
    Carburettor air-intake heat control 25

OTHER CONTROLS
    Intercommunication 26

PART II—HANDLING

Management of fuel system 27
Preliminaries 28
Starting engines and warming up 29
Testing engines and installations.. 30
Check list before taxying 31
Check list before take-off.. 32
Taking off 33
Climbing 34
General flying 35
Stalling 36
Diving 37
Check list before landing 38
Approach speeds 39
Mislanding 40
Beam approach 41
After landing 42

PART III—OPERATING DATA

Engine data—Merlin XX.. 43
Engine data—Merlin 22, 28 or 38 44
Engine data—Merlin 24 45
Flying limitations 46
Position error corrections 47
Maximum performance 48
Maximum range 49
Fuel capacity and consumptions 50

PART IV—EMERGENCIES

Engine failure during take-off 51
Engine failure in flight 52
Feathering 53
Unfeathering 54
Damage by enemy action 55
Undercarriage emergency operation 56

PART IV—*continued*   *Para.*

Flaps emergency operation	57
Bomb doors emergency operation	58
Bomb jettisoning	59
Fuel jettisoning	60
Parachute exits	61
Crash exits	62
Dinghy and ditching	63
Engine fire-extinguishers	64
Hand fire-extinguishers	65
Signal pistol	66
Signal cartridge stowage	67
Parachutes	68
Static line for parachuting wounded men	69
Emergency packs	70
Projectible kite container	71
Crash axes	72
Incendiary bombs	73
First-aid equipment	74

PART V—SUPPLEMENTARY NOTES FOR FLIGHT ENGINEER

Oil system	75
Coolant system	76
Hydraulic system	77
Pneumatic system	78
Electrical system	79
Undercarriage failure	80
Gauges	81

PART VI—ILLUSTRATIONS AND LOCATION OF CONTROLS

Location of controls	Page 46
Instrument panel	*Fig.* 1
Port side of cockpit	,, 2
Flight engineer's panel, Lancasters I & III	,, 3
Flight engineer's panel, Lancaster X	,, 4
Simplified fuel system diagram—Pulsometer pumps	,, 5
Simplified fuel system diagram—Immersed pumps	,, 6

AIR PUBLICATION 2062A, C & F—P.N.
Pilot's and Flight Engineer's Notes

PART I
DESCRIPTIVE

INTRODUCTION
1. The Lancaster I, III and X are heavy bombers, the difference between them lying mainly in the power plants; the Lancaster I is fitted with Merlin XX, 22 or 24 engines which have SU carburettors; the Lancaster III and X are fitted with Merlin 28 or 38 engines which have Bendix Stromberg pressure-injection carburettors; Hydromatic propellers are fitted to all. Lancaster X are Canadian built and differ from British Lancasters in some of the instruments and in the electrical system.

FUEL AND OIL SYSTEMS
2. **Fuel tanks.**—Three self-sealing tanks are fitted in each wing, numbered 1, 2 and 3 outboard of the fuselage between the front and rear spars. On some aircraft the tanks may be marked Inner, Centre and Outer instead of Nos. 1, 2 and 3. The positions are:

No. 1 (Inner): Between the fuselage and the inner engines
No. 2 (Centre): Between the inner and outer engines
No. 3 (Outer): Outboard of the outer engines.
Capacities are:
    Port and starboard No. 1: 580 gallons each
      ,, ,, ,,   No. 2: 383 ,,   ,,
      ,, ,, ,,   No. 3: <u>114</u> ,,   ,,
                        1,077 ,,   ,, side
                  or 2,154 ,, in all.

Provision is made on some aircraft for carrying one or two 400-gallon tanks fitted in the bomb cells; these tanks are connected so that their contents may be transferred into either or both No. 1 wing tanks and thence to the engines. When the maximum bomb load is carried, the No. 2 tanks should be filled first, and the remainder of the fuel put in No. 1 tanks. This is on account of strength considerations of the aircraft structure.

PART I—DESCRIPTIVE

3. **Fuel cocks.**—The pilot controls four master engine cocks (24, 30). On Lancaster I aircraft, the master engine cocks also control the slow-running cut-outs. The flight engineer controls two tank selector cocks (77) which select No. 1 or No. 2 tank on each side. (No. 3 tank replenishes No. 2, see below). A cross-feed cock (marked BALANCE COCK) connects the port and starboard supply systems, and is on the floor just forward of the front spar, with the handle visible through a hole in the spar cover.
When the 400-gallon tanks are fitted in the bomb cells they each have an ON-OFF cock situated behind the front spar in the centre of the fuselage.

4. **Vapour vent system** (Lancaster III and X aircraft only).—A vent pipe from each carburettor is connected to the No. 2 tank on the same side of the aircraft, and allows vapour and a small quantity of fuel (approx. $\frac{1}{2}$ gal. per hour, per carburettor, but some later carburettors may have a second vent allowing 10 gallons per hour) to return to the tank. This carburettor is designed to work full of fuel, and it therefore requires the vent to carry away any petrol vapour and dissolved air. It also assists in re-establishing the flow of fuel to the carburettors when the pipe-lines and pump have been run dry due to a tank emptying.

5. **Electric fuel booster pumps**
(i) Originally, on Lancaster I aircraft, immersed pumps were fitted in all tanks; Mod. 594 (temporary) removed the immersed pumps from No. 1 tanks and fitted stack pipes in their places.
A later Mod. 512 put back the immersed pumps in No. 1 and No. 2 tanks and incorporated suction by-pass lines to allow fuel to be drawn from the tanks when the pumps are not in use. In aircraft incorporating Mod. 539, including all Lancaster III aircraft, a Pulsometer FB Mk. I pump is fitted in each tank and by-pass lines are incorporated at No. 1 and No. 2 tanks. On Lancaster X aircraft, Thomson pumps, similar to the Pulsometer pumps, and by-pass lines are fitted.

PART I—DESCRIPTIVE

No. 3 tank is used to replenish No. 2 tank (*see* Figs. 5 & 6) by switching on the No. 3 tank pump. When the 400-gallon tanks are fitted in the bomb cells they each have a similar pump fitted to transfer their contents to the No. 1 tanks.

(ii) The main use of the electric fuel pumps in No. 1 and No. 2 tanks is to maintain fuel pressure at altitudes of approximately 17,000 ft. and over in temperate climates, but they are also used for raising the fuel pressure before starting and to assist in re-starting an engine during flight. If one engine fails during take-off and the electric fuel pump is not ON, air may be drawn back into the main fuel system before the master engine cock of the failed engine can be closed, thus causing the failure of the other engine on the same side; therefore at take-off the pumps in Nos. 1 and 2 tanks must be switched on; this is also a precaution against fuel failure during take-off as an immediate supply is available by changing over the tank selector cock. The pump in each tank in use should also be switched on at any time when a drop in fuel pressure is indicated or when it is necessary to run all engines from one tank by opening the cross-feed cock.

6. **Fuel contents gauges.**—On Lancasters I and III, the switch (76) on the flight engineer's panel must be set ON before the fuel contents gauges will indicate. On Lancaster X there is no fuel contents gauge switch; the gauges will indicate whenever electrical power is available.

7. **Fuel pressure indicators**—Fuel pressure warning lights (79) show when the fuel pressure at the carburettor falls below 6 lb./sq.in. on Lancaster I aircraft, and 10 lb./sq.in. on Lancaster III aircraft. They are switched off by the fuel contents gauges switch (76), and this switch must, therefore, always be on in flight. On Lancaster X, fuel pressure gauges (73) are fitted on the flight engineer's panel. They will indicate whenever battery power is available.

8. **Priming pumps.**—There is one cylinder priming pump in each inboard engine nacelle. Each pump serves one inboard and one outboard engine. On Lancasters I and III,

PART I—DESCRIPTIVE

two priming cocks are fitted in each nacelle. On Lancaster X the priming pump handle is turned to the left to prime the left engine, to the right to prime the right engine, and to the mid-position for off.

9. **Oil tanks.**—Each engine has its own tank; capacity $37\frac{1}{2}$ gallons of oil with $4\frac{1}{2}$ gallons air space.

0. **Oil dilution.**—The four push-buttons (81) are on the flight engineer's panel.

AIN SERVICES

1. **Hydraulic system**

(i) Each turret is operated by an individual engine-driven pump.

    Starboard outer engine: Mid upper turret
    Starboard inner ,, Front turret
    Port inner ,, Mid under turret, if fitted
    Port outer ,, Tail turret

(ii) Two pumps (one on each inboard engine), with a hand-pump as an alternative, charge a small accumulator and operate:

    Undercarriage
    Flaps
    Bomb doors
    Carburettor air intake shutters
    Fuel jettisoning

Owing to the large capacity of the flap and undercarriage jacks, it is not normally possible to operate them by the hand-pump in the time available in an emergency.

2. **Pneumatic system**

(i) A Heywood compressor on the starboard inboard engine charges an air bottle to 300 lb./sq.in. and operates:

    Wheel brakes
    Radiator shutters
    Supercharger rams (on Lancaster III and X and later Lancaster I aircraft)
    Idle cut-off rams (on Lancaster III and X aircraft only)

PART I—DESCRIPTIVE

A pressure-maintaining valve in the supply line from the air bottle only allows pressure to be supplied to the radiator shutters, superchargers, and idle cut-off rams, if the pressure in the air bottle exceeds 130 lb./sq.in. (This is to ensure sufficient pressure for the brakes, which operate at 80 lb./sq.in.) It is necessary, therefore, to check on the triple pressure gauge that pressure is sufficient before S ratio is engaged or the idle cut-off controls are operated.

(ii) A vacuum pump is fitted on each inboard engine, one for operating the instruments on the instrument flying panel, and the other for operating the gyros of the Mark XIV bombsight; the change-over cock (17) is on the right of the instrument panel beside the suction gauge (23), and in the event of failure of the vacuum pump supplying the flying instruments the changeover cock can be used to connect the serviceable pump with the flying instruments and cut out the bombsight. It is not possible to operate flying instruments and bombsight on one vacuum pump.

(iii) An RAE compressor fitted on the port inboard engine operates the Mark IV automatic controls and the computer unit of the Mark XIV bombsight. For operation of the Mark XIV bombsight, the automatic control cock must be set to OUT.

13. **Electrical system.**—Two 1,500-watt generators (one on each inboard engine), connected in parallel, charge the aircraft batteries (24 volt), and supply the usual lighting and other services, including

    Propeller feathering pumps
    Flap and undercarriage indicators
    Pressure head heating
    Fuel booster pumps
    Radio equipment
    Landing lamps
    Engine starting and booster coils
    Dinghy inflation
    Controls for radiator shutters rams, supercharger gear change rams, and idle cut-off rams (on Lancaster III and X aircraft)

PART I—DESCRIPTIVE

  Bomb gear and bombsight
  Fire-extinguishers
  Fuel contents gauges
  Fuel pressure warning lights or gauges
  Camera
  Heated clothing
  DR compass

An alternator may be fitted to each outboard engine, to supply any special radio equipment.

A ground/flight switch on the starboard side of the fuselage, immediately aft of the front spar, isolates the aircraft batteries when the aircraft is parked or when using a ground starter battery.

Two generator switches are provided on the electrical control panel.

AIRCRAFT CONTROLS

14. **Trimming tabs.**—The elevator (62), rudder (63) and aileron (61) tab controls (on the right of the pilot's seat) all operate in the natural sense and each has an indicator.

15. **Undercarriage control.**—The undercarriage lever (64) is locked in the DOWN position by a safety bolt (65) which has to be held aside in order to raise the lever. The bolt engages automatically when the lever is set down. The undercarriage may be lowered in an emergency by compressed air (*see* Part IV, para. 56).

    WARNING.—There is no automatic lock to prevent the undercarriage being raised by mistake when the aircraft is on the ground.

16. **Undercarriage indicator** (39).—On Lancasters I and III the indicators show as follows:

    Undercarriage locked down: Two green lights
    ,,   unlocked: Two red lights
    ,,   locked up: No lights

    The indicator switch (4) is interlocked so that it must be on when the port engine ignition switches are on. An auxiliary set of green lights can be brought into operation by pressing the central knob if failure of the main set is suspected. The red lamps are duplicated so that failure of one lamp does not affect the indication of undercarriage

PART I—DESCRIPTIVE

unlocked. The lights can be dimmed by turning the central knob. On Lancaster X a pictorial type of indicator is fitted. When the indicator switch is on and electrical power is available, the pictorial indicator shows the position of the undercarriage wheels and wing flaps at all times. The disappearance of small red flags shows when the wheels are locked up or down.

17. **Undercarriage warning horn.**—The horn sounds if either inboard throttle is closed when the undercarriage is not locked down. The outboard throttles do not operate the horn. A testing pushbutton and lamp are behind the pilot's seat, on the cockpit port rail.

18. **Flaps control.**—The push-pull handle (60) should be moved to the neutral position (located by a spring-loaded catch) after each flap movement. On Lancasters I and III the flap indicator (26) is switched on by a separate switch (27). If the flaps have been selected partly down, and it is desired to lower them fully, it may be found that the flaps will not lower further for some considerable time. This is due to the pressure in the accumulator having fallen below the pressure required to operate the flaps, but not sufficiently to cause the hydraulic pumps to cut in. To overcome this, move the flaps selector to UP, and then immediately put it fully DOWN; this causes the hydraulic pumps to cut in. In an emergency the flaps may be lowered by compressed air after lowering the undercarriage (*see* Part IV, para. 57).

19. **Bomb doors.**—The control (43) has two positions only. The bomb release system is rendered operative soon after the doors begin to open and before they are fully open. The position of the doors must therefore be checked visually before releasing bombs. If the bomb doors open only part way and then stop, it is probably due to icing around the hinges and joints, which raises the hydraulic pressure sufficiently to bring the cut-out into operation, which stops any further movement of the doors. If the bomb doors selector is moved to SHUT and then immediately to OPEN, the doors will usually open further; it may be necessary to repeat this several times to get the doors fully open.

PART I—DESCRIPTIVE

As strenuous pumping for 15 minutes is required to open the doors with engines stopped, they should be opened before stopping engines if the aircraft is to be bombed up before the next flight.

For emergency operation of bomb doors by compressed air, *see* Part IV, para. 58.

ENGINE CONTROLS

20. **Throttle controls**

(i) *Merlin XX, 22, 28 and 38 engines.*—Climbing boost +9 lb./sq.in. is obtained with the throttle levers (28) at the gate. On Merlin XX installations, and originally on Merlin 22, 28 or 38 installations, going through the gate gives a boost of +12 lb./sq.in. at ground level only. A later modification to Merlin 22, 28 or 38 engines gives +14 lb./sq.in. boost at ground level only with the throttle levers through the gate.

The boost control cut-out (32) gives +14 lb./sq.in. in M gear and +16 lb./sq.in. in S gear on all the above Merlins.

(ii) *Merlin 24 engines.*—Originally no boost control cut-out was fitted and no climbing gate. The fully forward position of the throttle lever gave +18 lb./sq.in. boost for take-off and combat. On these installations the automatic boost control does not allow the butterfly to open fully to maintain +9 lb./sq.in. boost up to full throttle height unless the throttle levers are progressively advanced to the fully forward position during the climb. However, a modification is now being introduced which overcomes this, and when this is incorporated a boost control cut-out is fitted which gives +18 lb./sq.in. for take-off and combat; the throttle quadrant is fitted with a gate at +9 lb./sq.in. boost, and the fully forward position gives + 14 lb./sq.in. at ground level only, for take-off at moderate loads.

(iii) On all Merlins, when climbing with a boost setting less than +9 lb./sq.in. the automatic boost control cannot open the throttle valve fully and the boost will begin to fall off before full throttle height is reached; the throttle lever should then be progressively advanced to maintain boost.

PART I—DESCRIPTIVE

21. **Mixture control**

(i) *Merlin XX, 22 and 24 engines (Lancaster I)*.—S.U. carburettors are fitted. The mixture strength is automatically controlled by boost pressure, and the pilot has no separate mixture control. A weak mixture is obtained below +7 lb./sq.in. boost (+4 lb./sq.in on Merlin XX). The carburettor slow-running cut-outs are operated by closing the master engine cocks.

(ii) *Merlin 28 and 38 engines (Lancasters III and X)*.—Bendix-Stromberg pressure injection carburettors are fitted. There is no pilot's mixture control, the mixture strength being regulated by the power so that a weak mixture is obtained below +7 lb./sq.in. and 2,650 r.p.m. The carburettor idle cut-outs, which are used in starting, and for stopping the engines, are operated by electro-pneumatic rams controlled by four SLOW-RUNNING CUT-OUT SWITCHES (11) on the pilot's panel just above the engine starter buttons. These switches have each two positions, the top one being the engine RUNNING position and the bottom one the IDLE CUT-OFF position. In the case of electrical or pneumatic failure the rams will return to the RUNNING position.

22. **Propeller controls.**—The speed control levers (29) for the Hydromatic propellers vary the governed r.p.m. from 3,000 down to 1,800. The feathering buttons (19) are on the right of the instrument panel. For feathering and unfeathering procedure *see* Part IV, paras. 53, 54.

23. **Supercharger controls**

(i) On early Lancaster I aircraft the supercharger controls for all four engines are operated mechanically by one lever.

(ii) On later Lancaster I and on all Lancaster III and X aircraft, the superchargers are operated by electro-pneumatic rams of the single-action spring-return type. In the case of electrical or pneumatic failure the rams will return to the M ratio position. A switch, fitted to the pilot's instrument panel immediately below the engine speed indicators, controls all four engines simultaneously, and a red warning light beside it (25) indicates S ratio on the ground only (i.e. when the undercarriage is down).

PART I—DESCRIPTIVE

24. **Radiator shutters.**—The shutters are automatically controlled when the switches forward of the flight engineer's panel are in the up position. When the switches are down, the thermostatic control is over-ridden, and the shutters are opened; this position should be used for all ground running, taxying and marshalling.

25. **Carburettor air-intake heat control.**—A single lever for the hydraulic operation of all four carburettors' hot air intakes is provided beside the pilot's seat. Hot air should not be used unless the air intakes become iced up, and as ice guards are fitted, this should rarely be necessary.

OTHER CONTROLS

26. **Intercommunication.**—On Lancaster X, Bendix interphone station boxes are fitted. For inter-communication the selector switch must be set to INTER and the INCREASE OUTPUT control set full on.

AIR PUBLICATION 2062A, C & F—P.N.
Pilot's and Flight Engineer's Notes

PART II

HANDLING

NOTE.—All speeds quoted are for aircraft with the Pilot's A.S.I. connected to the static vent (*see* para. 47). Speeds given in brackets are for aircraft on which the change to this connection has not been made.

27. **Management of fuel system**

(i) *Testing electric fuel booster pumps.*—Before starting the engines each booster pump should be tested by ammeter (most aircraft have a permanent ammeter fitted on the flight engineer's panel while some early aircraft may have an ammeter test socket into which the ammeter must be plugged); to do this, the switch for each pump should in turn be set to the up (test) position, after ensuring that the idle cut-off switches are in the IDLE CUT-OFF position and air pressure is greater than 130 lb./sq.in. (on Lancaster III and X aircraft); the ammeter reading should be perfectly steady and should be between 4 and 7 amps for a Pulsometer FB Mark I pump, between 3 and 5 amps for a Thomson pump, or between 2 and 4 amps for an immersed pump. Aircraft with Pulsometer pumps may be recognised by the small blisters on the underside of the wings.

(ii) *Use of tanks:*

(*a*) *Starting, warm-up and take-off.*—No. 2 tanks should be used first because the carburettor vents (*see* Part I, para. 4) return to these tanks, also because the contents of No. 1 tanks only can be jettisoned.

The electric fuel booster pumps in Nos. 1 and 2 tanks must be switched on for take-off, so that if for any reason the fuel supply from No. 2 tanks should fail, fuel pressure will be available immediately on turning the tank selectors to No. 1 tanks. *See* also Part I, para. 5.

(*b*) *In flight.*—After climbing to 2,000 feet, switch off booster pumps. Continue running on No. 2 tanks until 200 gallons remains in each, then change over to No. 1

PART II—HANDLING

tanks. Switch on No. 3 booster pumps to transfer their fuel to No. 2 tanks. Switch off the pumps when No. 3 tanks are empty. If overload fuel tanks are carried, continue to fly on No. 1 tanks until enough fuel has been used from them to enable the contents of the fuselage tanks to be transferred to them. Then change over to No. 2 tanks; turn on both long range fuel cocks (behind front spar) and switch on overload tank pump switches and fuel contents gauge. Transfer of fuel from long range tanks takes approximately 1 hour. Watch No. 1 fuel contents gauges and turn off each long range cock as soon as its respective No. 1 tank is filled. Turn off pumps when both No. 1 tanks are filled.

When over enemy territory, keep contents of Nos. 1 and 2 tanks approximately the same by running on each alternately for about half-an-hour.

(iii) *Use of cross-feed cock.*—The cock should be closed at all times, unless it is necessary in an emergency to feed fuel from the tanks in one wing to the engines in the other wing. If the cross-feed cock is open for this purpose, only one tank should be turned on to feed all working engines, and the pump in this tank should be switched on.

(iv) *Changing fuel tanks on Lancaster III and X aircraft.*—If the engine cuts owing to exhaustion of fuel in one tank, back-firing may occur on turning on to another tank. For when the tank empties, the fuel pressure drops, and when the pressure falls to 4 lb./sq.in. fuel injection ceases. When the new tank is turned on, the carburettor restarts and delivers the fuel already in it, but this supply is followed by vapour from the fuel pipelines, causing weak mixture and back-firing until fuel is delivered from the new tank.

When an engine cuts due to exhaustion of one tank:

(*a*) Close the throttle and change over to another tank.

(*b*) Idle the engine till it runs smoothly and open up slowly.

(*c*) The use of the booster pump in the tank turned on will help to restart the engine.

PART II—HANDLING

28. Preliminaries

(i) *Before entering aircraft.*—Check pitot head covers removed.
Check all cowling and inspection panels, and leading edge secured. Check tyres for creep.

(ii) *On entering aircraft.*—Check security of emergency escape hatches
Check emergency air bottle pressure (1,100–1,200 lb./sq.in.)
Check hydraulic accumulator pressure (220 lb./sq.in. minimum, under no hydraulic pressure).
Check fuel cross-feed cock OFF.
Turn ground/flight switch to FLIGHT.
Switch on undercarriage indicator and flaps indicator switches (if fitted) and check indicators.
Switch on fuel contents gauges switch (if fitted) and leave it on, and check fuel contents.
Check master engine cocks OFF.

29. Starting engines and warming up

(i) Test fuel booster pumps by ammeter (*see* para. 27).
On Lancasters III and X aircraft the fuel tank booster pump must *never* be switched on with the master engine cock open and the engine stationary, unless the slow-running cut-out switch is in the IDLE CUT-OFF (down) position and the air supply pressure not less than 130 lb./sq.in.

(ii) Turn ground/flight switch to GROUND and plug in ground starter battery.

(iii) Set engine controls as follows:

Master engine cocks	OFF
Throttles	$\frac{1}{2}$ in. open
Propeller controls	Fully up
Slow-running cut-out switches	IDLE CUT-OFF
Supercharger control	M ratio (Warning light not showing)
Air intake heat control	COLD
Radiator shutters	Over-ride switches at AUTOMATIC

PART II—HANDLING

(iv) Have a fire-extinguisher ready in case of emergency.

(v) Turn tank selector cock to No. 2 tank (*see* para. 27 (ii)) and turn on the master engine cock of the engine to be started. On Lancasters III and X aircraft the master engine cocks of all other engines which are not running must be OFF.

(vi) Switch on the booster pump in the No. 2 tank to be used.

(vii) High-volatility fuel (stores ref. 34A/111) should be used, if outside priming connection is fitted, for priming at air temperatures below freezing. The ground crew will work the priming pump until the fuel reaches the priming nozzles; this may be judged by an increase in resistance.

(viii) Switch on the ignition and booster coil, and press the starter button. Turning periods must not exceed 20 seconds with a 30-second wait between each. The ground crew will work the priming pump as firmly as possible while the engine is being turned; it should start after the following number of strokes, if cold:

Air temp., °C.	+30	+20	+10	0	−10	−20
Normal fuel ..	3	4	7	12		
High-volatility fuel ..				4	8	18

It will probably be necessary to continue priming after the engine has fired and until it picks up on the carburettor.

(ix) *On Lancaster III and X aircraft.*—As soon as the engine fires regularly, move the slow-running cut-out control to the ENGINE RUNNING (up) position. If, however, the engine shows signs of being over-rich, return the slow-running cut-out control to the IDLE CUT-OFF (down) position, until it is running smoothly.

(x) When all the engines are running satisfactorily, switch off the booster-coil switch. The ground crew will screw down the priming pumps and turn off the priming cocks (if fitted).

(xi) Have the ground/flight switch turned to FLIGHT and the ground battery removed.

(xii) Open each engine up slowly to 1,200 r.p.m. and warm up at this speed.

(xiii) Switch DR compass ON and SETTING.

PART II—HANDLING

30. **Testing engines and installations**
    While warming up :
    (i) Check temperatures and pressures, and test operation of hydraulic system by lowering and raising flaps and bomb doors—but do not test bomb doors if a bomb load is on board.
    (ii) Switch off electric fuel booster pump so as to test engine-driven pumps.
    After warming up :
    NOTE.—The following comprehensive checks should be carried out after repair, inspection other than daily, or otherwise at the pilot's discretion. Normally they may be reduced in accordance with local instructions.
    (iii) Switch radiator shutters over-ride switches to OPEN.
    (iv) At 1,500 r.p.m. test each magneto in turn to ensure that no magnetos are unserviceable.
    (v) Open up to +4 lb./sq.in. boost and check operation of two-speed supercharger. R.p.m. should fall when S ratio is engaged, and, on aircraft with electro-pneumatically operated supercharger gear change, red warning light should come on. Return to M ratio.
    (vi) At the same boost, check operation of constant-speed propeller. R.p.m. should fall to 1,800 with the control fully down
    (vii) Open throttle to the take-off position and check take-off boost and r.p.m.—*see* Part I, para. 20.
    (viii) Throttle back to +9 lb./sq.in. boost; check that r.p.m. fall below 3,000 and if not throttle back until a drop is shown, to ensure that the propeller is not constant-speeding. Then test each magneto in turn. The drop should not exceed 100 r.p.m.

31. **Check list before taxying**
    Ground/flight switch FLIGHT
    Navigation lights .. On if required
    Altimeter .. .. Set
    Instrument flying panel Check vacuum on each pump
    $-4\frac{1}{2}$ lb./sq.in.
    Radiator shutter
        switches OPEN
    Brakes pressure .. Supply 250–300 lb./sq.in.

PART II—HANDLING

32. **Check list before take-off**

Auto controls—	Clutch	IN
	Cock	OUT
DR compass ..	..	NORMAL
Pitot head heater switch		ON
T —Trimming tabs ..		Elevator slightly forward
		Rudder neutral
		Aileron neutral
P —Propeller controls		Fully up
F —Fuel ..	..	Check contents of tanks
		Master engine cocks ON
		Tank selector cocks to No. 2 tanks
		Crossfeed cock OFF
		Booster pumps in Nos. 1 and 2 tanks ON
Superchargers	..	MOD
Air intake	..	COLD
Radiator shutters switches		AUTOMATIC
F —Flaps ..	..	15°–20° down

33. **Taking off**
    (i) Open the throttles to about zero boost against the brakes to see that engines are responding evenly. Throttle back, release brakes, and open throttles gently checking the tendency to swing to port by advancing port throttles slightly ahead. This will give as good a take off as taking off against the brakes, and renders it easier to correct swing.
    (ii) The tail should be raised as quickly as possible after the throttles are fully open and the aircraft eased off the ground at not less than 95(90) m.p.h. I.A.S. if loaded to 50,000 lb., or 105(100) m.p.h. I.A.S. if loaded to 60,000 lb.
    (iii) Safety speed is 130(125) m.p.h. I.A.S.
    (iv) Raise the flaps at a safe height, not below 500 feet when heavily loaded, and return selector to neutral. Raising flaps causes a nose-down change of trim.
    (v) Switch off electric fuel booster pumps of Nos. 1 and 2 tanks after initial climb, but if a warning light comes on, or fuel pressure drops below 10 lb./sq.in., switch on No. 2 pumps immediately.

PART II—HANDLING

34. Climbing

(i) The recommended speed for a quick climb is 160(155) m.p.h. I.A.S. The most comfortable climbing speed is about 175(170) m.p.h. I.A.S.

(ii) Switch on electric fuel pumps of tanks in use, at any signs of fuel starvation (at approximately 17,000 feet in temperate climates).

35. General flying

(i) *Stability.*—At normal loadings and speeds, stability is satisfactory.

(ii) *Controls.*—The elevators are relatively light and effective, but tend to become heavy in turns. The ailerons are light and effective, but become heavy at speeds over 260 m.p.h. I.A.S.
The rudders also become heavy at high speeds.

(iii) *Change of trim:*

Undercarriage UP	Slightly nose up
Undercarriage DOWN	Slightly nose down
Flaps up to 25° from fully DOWN	Slightly nose down
Flaps up from 25°	Strongly nose down
Flaps down to 25°	Strongly nose up
Flaps fully DOWN from 25° ..	Slightly nose up
Bomb doors open	Slightly nose up

(iv) *Flying at low airspeeds.*—Flaps may be lowered about 15°–20°, r.p.m. set to 2,650, and the speed reduced to about 130(125) m.p.h. I.A.S.

36. Stalling

(i) Just before the stall, slight tail buffeting occurs.

(ii) There is no tendency for a wing to drop.

(iii) The stalling speeds in m.p.h. I.A.S. at 50,000 lb. are:
    Undercarriage and flaps up   110 (95)
    Undercarriage and flaps down 92 (83)

PART II—HANDLING

37. Diving

(i) The aircraft becomes increasingly nose heavy in a high-speed dive. The elevator tab control should not be used to help the entry into the dive, but it should be used to trim out the pull necessary in the later stages of the dive.

(ii) The flight engineer should be ready to assist the pilot as required.

38. Check list before landing

Auto-pilot control cock	OUT (clutches may be left IN)
Superchargers	M (low) ratio
Air intake	COLD
Brake pressure	Supply pressure 250–300 lb./sq.in.

Reduce speed to below 200 m.p.h. I.A.S. and carry out the following drill:

Flaps	20° down on circuit
U —Undercarriage	DOWN (check by indicator, visually, and horn)
P —Propeller	Controls up to at least 2,850 r.p.m.
F —Flaps	DOWN on final approach (handle neutral) (*see* para. 18)
F —Fuel	Booster pumps ON in tanks in use

39. Approach speeds

Recommended speeds for the approach in m.p.h. I.A.S.:

	45,000 lb.	55,000 lb.
Engine-assisted	110(100)	120(110)
Glide	120(110)	130(120)

40. Mislanding

(i) The aircraft will climb satisfactorily with the undercarriage and flaps down.

(ii) Climb at about 140(130) m.p.h. I.A.S. and, after raising the undercarriage, start raising the flaps a little at a time, retrimming as necessary.

PART II—HANDLING

41. Beam approach

	Indicated height: feet	I.A.S. m.p.h.	R.p.m.	Approx. boost: lb./sq.in.	Actions	Change of trim
Preliminary approach	1,500	135	2,400	−1	Set 25° flap for all manoeuvring	Nose up strongly
On QDR +30°	1,500	135	2,650	+2	Undercarriage down	Nose down slightly
At outer marker	600	130	2,650	−2 to 0 (descent at 400 ft./min.) +2 (level flight)		
At inner marker	150	120	2,850	0 to +2	Flaps fully down	Nose up slightly
Overshoot	up to 300	130–140	2,850	+9	Raise flaps to 25° Raise undercarriage	Nose down slightly Nose up slightly

All four throttles should be used together throughout the approach.
An increase in the rate of sink has no effect on the controls.

PART II—HANDLING

42. After landing

(i) Before taxying, raise the flaps and open the radiator shutters.

(ii) The outer engines may be stopped and taxying done on the inners. This is preferable to stopping the inner engines, as the brakes compressor is on the starboard inner engine, and the outer engines are more liable to overheat.

(iii) Before stopping the engines, open the bomb doors for bombing up (if required).

(iv) Switch off all booster pumps before stopping engines.

(v) *Stopping engines, Merlin XX, 22, 24.* With the engines running at 800 r.p.m., turn OFF the master engine cocks and switch OFF the ignition after the engines have stopped.

(vi) *Stopping engines, Merlin 28 or 38.*—To stop an engine check that the air pressure gauge reads at least 130 lb./sq.in. (*see* para 12 (i)) (if not, open up starboard inner engine which drives the compressor) and move the slow-running cut-out switch to the IDLE CUT-OFF (down) position with the engine running at about 800 r.p.m. Then switch off the ignition and leave the slow-running cut-out switch in the IDLE CUT-OFF (down) position.

> NOTE.—Merlin 28 and 38 engines must not be stopped by turning off the master engine cock, as this will empty the carburettor of fuel and fill it with air. This entails much trouble when starting the engine again.

When all engines have been stopped, turn off master engine cocks, check again that all booster pumps are OFF, and then return the slow-running cut-outs to the ENGINE RUNNING position.

> NOTE.—If the switches are left in the IDLE CUT-OFF position, the rams will return to the ENGINE RUNNING position as soon as the master switch is turned to GROUND. Then, if any ground maintenance work is carried out which necessitates turning on and off the master switch, the rams are being continually operated.

PART II—HANDLING

(vii) Switch off all electrical switches and turn master electrical switch to GROUND.

(viii) *Oil dilution.*—See *Pilot's Notes General A.P.* 2095. The correct dilution period for this aircraft is:

    Air temperatures above −10°C. .. One minute
       „   „   below −10°C. .. Two minutes

AIR PUBLICATION 2062A, C & F—P.N.
Pilot's and Flight Engineer's Notes

PART III
OPERATING DATA

43. Engine data—Merlin XX

(i) *Fuel*—100 octane only.

(ii) *Oil*—See A.P.1464/C.37

(iii) *Engine limitations* with 100 octane fuel:

		R.p.m.	Boost lb./sq.in.	Temp. °C. Coolant	Oil
MAX. TAKE-OFF TO 1,000 FEET	M	3,000	+12*		
MAX. CLIMBING 1 HOUR LIMIT	M / S	2,850	+ 9	125	90
MAX. RICH CONTINUOUS	M / S	2,650	+ 7	105	90
MAX. WEAK CONTINUOUS	M / S	2,650	+ 4	105	90
COMBAT 5 MINS. LIMIT	M	3,000	+14*	135	105
	S	3,000	+16*	135	105

*See Part I, para. 20.

OIL PRESSURE:
  NORMAL .. .. 60/80 lb/sq.in.
  MINIMUM .. .. 45 lb/sq.in.
MINIMUM TEMPS. FOR TAKE-OFF:
  OIL .. .. 15°C.
  COOLANT .. .. 60°C.

44. Engine data—Merlin 22, 28 or 38

(i) *Fuel*—100 octane only.

(ii) *Oil*—See A.P.1464/C.37.

(iii) *Engine limitations* with 100 octane fuel:

		R.p.m.	Boost lb./sq.in.	Temp. °C. Coolant	Oil
MAX. TAKE-OFF TO 1,000 FEET	M	3,000	+14*		
MAX. CLIMBING 1 HOUR LIMIT	M / S	2,850	+ 9	125	90
MAX. CONTINUOUS	M / S	2,650	+ 7	105	90
COMBAT 5 MINS. LIMIT	M	3,000	+14*	135	105
	S	3,000	+16*	135	105

*See Part I, para 20.

OIL PRESSURE:
  NORMAL .. .. 60/80 lb./sq.in.
  MINIMUM .. .. 45 lb./sq.in.
MINIMUM TEMPS. FOR TAKE-OFF:
  OIL .. .. 15°C.
  COOLANT .. .. 60°C.

PART III—OPERATING DATA

45. Engine data—Merlin 24

(i) *Fuel*—100 octane only.

(ii) *Oil*—See A.P.1464/C.37.

(iii) *Engine limitations* with 100 octane fuel:

		R.p.m.	Boost lb./sq.in.	Temp. °C. Coolant	Oil
MAX. TAKE-OFF TO 1,000 FEET	M	3,000	+18*		
MAX. CLIMBING 1 HOUR LIMIT	M S	2,850	+ 9	125	90
MAX. CONTINUOUS	M S	2,650	+ 7	105	90
COMBAT 5 MINS. LIMIT	M S	3,000	+18*	135	105

* +18 lb./sq.in. boost must not be used below 2,850 r.p.m.

```
OIL PRESSURE:
    NORMAL  ..  ..  ..  ..  ..  60/80 lb./sq.in.
    MINIMUM ..  ..  ..  ..  ..  45 lb./sq.in.
MINIMUM TEMP. FOR TAKE-OFF:
    OIL     ..  ..  ..  ..  ..  ..  ..  15°C.
    COOLANT ..  ..  ..  ..  ..  ..  ..  60°C.
```

46. Flying limitations

(i) The aircraft is designed for manœuvres appropriate to a heavy bomber and care must be taken to avoid imposing excessive loads with the elevators in recovery from dives and in turns at high speed. Spinning and aerobatics are not permitted. Violent use of the rudder at high speeds should be avoided.

(ii) *Maximum speeds in m.p.h. I.A.S.:*

```
Diving              ..  ..  ..  360
Bomb doors open     ..  ..  as for diving
Undercarriage down  ..  ..  200
Flaps down  ..      ..  ..  200
```

PART III—OPERATING DATA

(iii) *Maximum weights:*

Take off and straight flying .. 65,000 lb., provided that the following mods. are incorporated:
Mod. 505 or 518
„ 588 or 598
„ 811 or SI/RDA.660
„ 1004
63,000 lb., if these mods. are not incorporated.

Landing and all forms of flying .. 55,000 lb.

Flying should be restricted to straight and level until weight is reduced to 63,000 lb.

(iv) *Bomb clearance angles:*

Dive .. .. .. 30°
Climb .. .. .. 20°
Bank .. .. .. 10° (with S.B.C. 25°)

47. Position error corrections

All handling speeds are quoted for aircraft with the pilot's A.S.I. connected to the static vent, in the port side of the fuselage. The position error for the static vent connection is -1 m.p.h. at all speeds from 140 m.p.h. I.A.S. upward.

The position error correction for aircraft on which the static vent connection has not been made is as follows:—

From	120	140	160	180	200	} m.p.h. I.A.S.
To	140	160	180	200	250	
Add	12	10	8	6	4	m.p.h.

With bomb doors open, the correction is 2 m.p.h. higher between 160 and 200 m.p.h. I.A.S.

PART III—OPERATING DATA

48. Maximum performance

(i) *Climbing:*

160 (155) m.p.h. I.A.S. to 12,000 ft.
155 (150) m.p.h. I.A.S. from 12,000 to 18,000 ft.
150 (145) m.p.h. I.A.S. from 18,000 to 22,000 ft.
145 (140) m.p.h. I.A.S. above 22,000 ft.

Change to S ratio when boost has fallen to +6 lb./sq.in.

(ii) *Combat.*—Use S ratio if the boost obtainable in M ratio is more than 3 lb./sq. in. (4 lb./sq.in. with Merlin 24) below maximum combat boost.

49. Maximum range

(i) *Climbing.*—160 (155) m.p.h. I.A.S. at +7 lb./sq.in. boost with Merlin 22, 24, 28 or 38, +4 lb./sq.in. with Merlin XX, and 2,650 r.p.m. Change to S ratio when maximum boost obtainable in M ratio has fallen by 3 lb./sq.in.

(ii) *Cruising* (including descent):

(a) Fly in M ratio at maximum obtainable boost not exceeding +4 lb./sq.in. with Merlin XX, +7 lb./sq.in. with Merlin 22, 24, 28 or 38, obtaining the recommended airspeed by reducing r.p.m. which may be as low as 1,800 if this will give the recommended speed. Higher speeds than those recommended may be used if obtainable in M ratio at the lowest possible r.p.m.

(b) The recommended speeds are:

Fully loaded (outward journey):

Up to 15,000 ft.,   170 (165) m.p.h. I.A.S.
At 20,000 ft. in S ratio, 160 (155) m.p.h. I.A.S.

Lightly loaded (homeward journey):

160 (155) m.p.h. I.A.S.

(c) Engage S ratio when the recommended speed cannot be maintained at 2,500 r.p.m. in M ratio.

(iii) The use of warm intakes will reduce air miles per gallon considerably. On this installation there is no need to use warm air unless intake icing is indicated by a drop of boost. *See* A.P. 2095—Pilot's Notes General.

PART III—OPERATING DATA

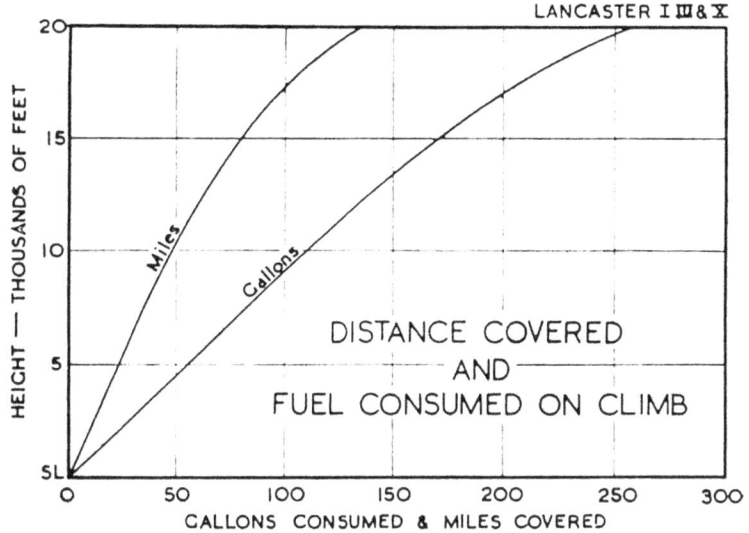

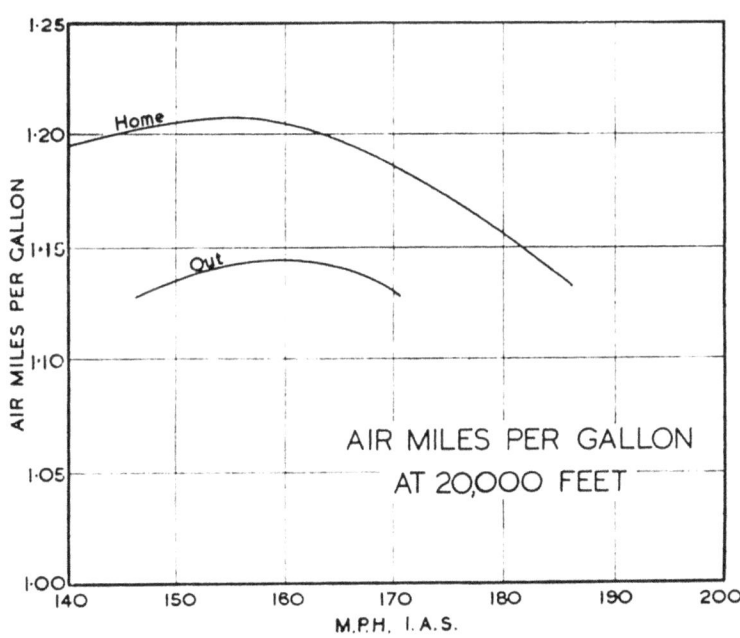

PART III—OPERATING DATA

50. Fuel capacity and consumptions

(i) Capacity: Two No. 1 tanks .. 1,160 gallons
Two No. 2 tanks .. 766 gallons
Two No. 3 tanks .. 228 gallons

Total .. 2,154 gallons

(ii) *Weak mixture consumptions, Merlin XX, 22 or 24:*
The following figures are the approximate total gallons per hour and apply in M ratio between 8,000 and 17,000 feet, and in S ratio between 14,000 and 25,000 feet.

Boost lb./sq.in.	R.P.M.		
	2,650	2,300	2,000
+7*	260*	225*	212*
+4	228	204	188
+2	212	188	172
0	192	172	150
−2	172	156	140
−4	152	136	124

* These figures do not apply to Merlin XX.

(iii) *Weak mixture consumptions Merlin 28 and 38 engines:*
The following figures are the approximate total gallons per hour for the aircraft and apply in M ratio at 5,000 ft. and S ratio at 15,000 ft. One gallon per hour should be added for every 1,000 ft. above these heights.

Boost lb./sq.in.	R.P.M.				
	2,650	2,400	2,200	2,000	1,800
+7	240	235	217	200	—
+4	216	204	196	180	—
+2	196	184	176	164	—
0	172	164	156	144	128
−2	148	140	128	124	112
−4	124	120	108	104	96

PART III—OPERATING DATA

(iv) *Rich mixture consumption, Merlin XX, 22, 24*:

Boost lb./sq.in.	R.p.m.	Total gallons per hour
+14	3,000	500
+12	3,000	460
+ 9	2,850	380
+ 7*	2,650*	320*

* Merlin XX only.

(v) *Rich mixture consumptions, Merlin 28, 38*:

Boost lb./sq.in.	R.p.m.	Total gallons per hour
+9	2,850	420

AIR PUBLICATION 2062A, C & F—P.N.
Pilot's and Flight Engineer's Notes

PART IV

EMERGENCIES

51. Engine failure during take-off

(i) If for any reason the booster pumps in the tanks being used are not ON, the master fuel cock of the failed engine must be turned off before feathering.

(ii) If one outer engine fails, the aircraft can be kept straight provided 130 (125) m.p.h. I.A.S. has been reached.

(iii) Climbing speed on three engines should be 135 (130) m.p.h. I.A.S. at moderate loads, or 140 (135) m.p.h. I.A.S. if heavily loaded.

(iv) As soon as the undercarriage is up, raise the flaps a little at a time, retrimming as necessary.

52. Engine failure in flight

(i) If an engine is stopped, air may be drawn into the fuel system through the carburettor of the failed engine. To prevent this, the master fuel cock must be turned off before feathering, unless the booster pump for the tank in use is on. If an engine seizes up, its master fuel cock should be turned off immediately to ensure that the fuel supply to the other engine on the same side is not affected.

(ii) If the failed engine cannot be made to pick up again, feather its propeller and switch off. On Lancaster III and X, set mixture control to IDLE CUT-OFF.

(iii) *Handling on three engines.*—The aircraft will maintain height at full load on any three engines at 10,000 feet, and can be trimmed to fly without foot load. Maintain at least 145 (140) m.p.h. I.A.S. The automatic pilot has sufficient power to maintain a straight course with either outboard engine out of action, but only if assisted by the rudder trimming tab.

PART IV — EMERGENCIES

(iv) *Landing on three engines.*—Lowering of flaps to 20° and of undercarriage may be carried out as normally on the circuit, but further lowering of the flaps should be left until final straight approach. The final approach should be made at 120–125 m.p.h. I.A.S. using as little power as possible, and rudder trim should not be wound off until definitely committed to landing. Use all good engines to regulate approach. (*See* A.P. 2095, Part IV, Note D.)

(v) *Handling on two engines.*—It should be possible to maintain height below 10,000 feet at 140 (135) m.p.h. I.A.S. on any two engines after release of bombs and with half fuel used; but with two engines dead on one side, the foot load will be very heavy. The automatic pilot will not cope with flight with two engines dead on one side.

(vi) *Landing on two engines.*—The circuit should be made with the good engines on the inside of the turn, and undercarriage and flaps left as late as practicable. Keep extra height in hand, if possible, and lower the undercarriage later than usual, but have it locked down just before the final approach. The approach should be made at 130–135 m.p.h. I.A.S. in a glide. When certain of getting into the airfield, lower flaps for landing. Do not wind off trim until final approach can be made in a glide, as some power may be necessary in the early stages. (*See* A.P. 2095, Part IV, Notes C and D.)

(vii) Do not attempt to maintain height above 10,000 feet, either on three or two engines.

(viii) *Fuel system.*—The cross-feed cock should only be turned on when it is desired to feed fuel from port (or starboard) tanks to starboard (or port) engines. In this case all live engines should be fed from one tank and the fuel booster pump for this tank should be on. The fuel selector for the tanks on the other side of the aircraft should be off. At all other times, the cross-feed cock should be off.

53. **Feathering**

(i) (*a*) *Practice feathering.*—Check that booster pump in the tank feeding the engine to be stopped is on. *See* para. 52 (i). Set propeller control to low r.p.m. and throttle to give a moderate cruising boost.

PART IV—EMERGENCIES

(b) *Emergency feathering.*—If possible, master fuel cock should be turned off immediately before feathering, but if this is not possible turn fuel cock off immediately after feathering. *See* para. 52 (i).

(ii) Hold the button in only long enough to ensure that it stays in by itself; then release it so that it can spring out when feathering is complete.
(iii) Close the throttle; on Lancaster III and X aircraft move the slow-running cut-out switch to IDLE CUT-OFF position.
(iv) Switch off only when the engine has stopped.
(v) Engine auxiliaries which will be affected by feathering:

Port outer	Alternator for special radio, rear turret hydraulic pump.
Port inner	Generator, main services hydraulic pump, mid-under turret hydraulic pump, R.A.E. compressor for Automatic Pilot Mark IV and computer unit of Mark XIV bombsight, No. 1 vacuum pump.
Starboard inner	Generator, main services hydraulic pump, front turret hydraulic pump, Heywood compressor for pneumatic system, No. 2 vacuum pump.
Starboard outer	Alternator for special radio, mid-upper turret hydraulic pump.

54. **Unfeathering**

NOTE.—It is preferable not to unfeather at speeds above normal cruising speed to avoid any risk of overspeeding.

(i) Set throttle closed or slightly open, propeller control fully down and ignition on.
(ii) Switch on the fuel booster pump of the tank in use, and turn on master engine cock.
(iii) Hold the button in and, on Lancaster III and X aircraft, as the engine starts turning set the slow-running cut-out switch to RUNNING (up) position. Continue to hold the button in until r.p.m. reach at least 1,500 and not more than 1,800.
(iv) If the propeller does not return to normal constant-speed operation, open throttle slightly.

PART IV—EMERGENCIES

55. Damage by enemy action

(i) *Fires.*—If fire occurs in bomb bay, pilot should open bomb doors, jettison bombs, and dive.

(ii) *Flight engineer's checks:*

Check fuel contents gauges. Should any abnormal consumptions be shown on the tank the aircraft is running on, or should any of the other tanks show a loss of fuel, proceed as follows:

(a) Cross-feed cock ON.

(b) Run all engines on the damaged tank, and switch on its booster pump. The fuel selector for the tanks on the other side of the aircraft should be off.

(c) When contents of damaged tank fall to 20 gallons, turn on fuel selector for the corresponding tank on the other side of the aircraft and turn OFF cross-feed valve.

(d) Watch fuel pressure warning lights or fuel pressure gauge and change over to other tank as soon as pressure drops; switch off booster pump in empty tank.

56. Undercarriage emergency operation

If the hydraulic system fails the undercarriage can be lowered by compressed air from a special bottle or bottles, irrespective of the position of the undercarriage lever.

NOTE.—The flap selector should be neutral before using the undercarriage emergency air system.

On early aircraft the control is just aft of the front spar, but on later aircraft the knob (80) for working the air system is just forward of the flight engineer's panel. The undercarriage cannot be raised again by this method. Although the undercarriage will lower by this method, irrespective of the position of the normal undercarriage selector, the latter should be left in the down position for and after landing; otherwise, any leakage of air pressure may cause the undercarriage locks to be released and the undercarriage to collapse.

PART IV—EMERGENCIES

57. Flaps emergency operation

After lowering the undercarriage by turning on the emergency air cock, the flaps may be lowered by operating the flaps control, which admits the air pressure to the flaps system. The flaps can be raised again, but there may not be sufficient air pressure to lower the flaps a second time; furthermore it may cause the header tank to burst. If it is absolutely necessary to raise the flaps by emergency method extreme care must be taken to raise them slowly by stages. If the flaps are lowered by the emergency method before landing, flaps must be left down after landing, owing to the likelihood of bursting the header tank.

58. Bomb doors emergency operation

On aircraft in which Mod. 757 is incorporated, the bomb doors can be operated in an emergency by compressed air from an air bottle (entirely separate from undercarriage and flaps emergency air bottle).

If, when pilot selects bomb doors open or closed, the doors fail to operate, the air bomber should withdraw the pin on the emergency air control in the nose, and hold the control down against the spring; when the doors have fully opened (or closed), the air bomber should release the control. There is sufficient pressure in the air bottle for three or four operations.

59. Bomb jettisoning

(i) Open bomb doors, and check visually that both are fully open. *See* para. 19.

(ii) Then jettison containers first by switch (15) on right of dash.

(iii) Jettison bombs by handle (16) beside container jettison switch.

(iv) Close bomb doors.

PART IV—EMERGENCIES

60. Fuel jettisoning

NOTE.—Contents of No. 1 tanks only can be jettisoned.

(i) Reduce speed to 150 (145) m.p.h. I.A.S. and lower flaps 15°.

(ii) Lift and turn jettison control on left of pilot's seat. Return control after jettisoning.

(iii) The jettison valve should be closed while there is still about 100 gallons remaining in each tank; if the jettison valve is left open, all the fuel will be jettisoned less approximately 70 gallons, but the last 30 gallons of jettisonable fuel runs out slowly and is inclined to get splashed over the fuselage. The jettison valve may be closed at any time during jettisoning.

Approximate weight of jettisonable fuel, leaving 100 gallons in each tank, is 6,900 lb.

61. Parachute exits

(i) Hatch in floor of nose should be used by all members of the crew if time is available; it is released by handle in centre, lifted inwards and jettisoned.

(ii) Main entrance door should be used as a parachute exit only in extreme emergency.

62. Crash exits

Three push-out panels are fitted in the roof, one above the pilot, one just forward of the rear spar, and one forward of the mid-upper turret.

63. Dinghy and ditching

(i) A type J dinghy stowed in the starboard wing may be released and inflated:

(a) from inside by pulling the release cord running along the fuselage roof aft of the rear spar;

(b) from outside by pulling the loop on the starboard side, rear of the tail plane leading edge.

(c) automatically by an immersion switch.

(ii) The flaps should be lowered 30° for ditching, but if the flaps will not lower by the hydraulic system, do not attempt to lower them by the compressed air system, as this will also cause the undercarriage to lower (*see* paras. 56 and 57).

PART IV—EMERGENCIES

64. **Engine fire-extinguishers**

    Each engine is provided with a fire-extinguisher bottle which can be operated by pushbuttons on the pilot's instrument panel; the fire-extinguishers are also operated automatically by a crash switch (on the starboard side of the nose) and, when the undercarriage is down, by a gravity switch (also on the starboard side of the nose).

    NOTE.—The automatic flame switches have been disnected. Therefore in flight the operation of the engine fire-extinguisher is left entirely to the pilot's discretion.

65. **Hand fire-extinguishers**

    One on port side of air bomber's compartment.
    One on port side of pilot's seat.
    One on starboard side aft of rear spar.
    One on starboard side forward of mid-upper turret.
    One at tail turret.

66. **Signal pistol.**—This is stowed on top of the front spar; the firing position is in the roof forward of the stowed position.

67. **Signal cartridge stowage.**—Starboard side of fuselage just forward of front spar.

68. **Parachutes**

    One in air bomber's compartment, on aft bulkhead.
    One behind pilot's seat.
    One on starboard wall at rest station.
    One on starboard wall forward of mid-upper turret.
    One on port side, opposite entrance door.
    One forward of tail turret.

69. **Static line for parachuting wounded men**

    (i) If possible, fly aircraft at 130 m.p.h. I.A.S. with 15° flap.
    (ii) Assist casualty to air bomber's compartment and place him feet first facing aft.
    (iii) Check casualty's parachute harness, fit parachute, remove helmet.

PART IV—EMERGENCIES

(iv) Remove static line from stowage which is situated on starboard side of front exit. Care should be taken that the threads keeping the static line folded up are not broken. Take snap hook at end of static line and attach to parachute as follows.

(v) Pass the safety becket on the static line through the double 8 cord loop, then pass the small snap hook through the safety becket.

(vi) Snap the hook down on to the rip-cord handle. Insert safety pin to lock the shroud of the snap hook.

(vii) Stow the slack of the static line between the becket and the snap hook under the adjacent pack elastic to prevent this slack length getting caught up on anything and thus pulling the rip-cord too soon.

(viii) Open and jettison front hatch.

(ix) Slide the man through the exit feet first facing aft. Care must be taken to keep his hands to his sides. Do not hold on to the static line by hand.

70. **Emergency packs.**—On starboard side at rest station.

71. **Projectible kite container.**—Along port side at rest station.

72. **Crash axes**

One on port side of fuselage aft of main entrance door. One on starboard wall in front of rear spar.

73. **Incendiary bombs.**—Two are provided on front face of front spar for destruction of aircraft.

74. **First-aid equipment.**—Starboard side of fuselage aft of main entrance door.

AIR PUBLICATION 2062A, C & F—P.N.
Pilot's and Flight Engineer's Notes

PART V

SUPPLEMENTARY NOTES FOR FLIGHT ENGINEER

75. **Oil system.**—A self-sealing oil tank is fitted in each nacelle; the normal capacity is $37\frac{1}{2}$ gallons with $4\frac{1}{2}$ gallons air space. The normal type of oil used is DTD. 472B. A stack pipe in each tank retains 2–3 gallons for feathering the propeller. Normal high-pressure oil feeds the propeller constant speed unit.
    Under cruising conditions, it is recommended that the oil temperature should not exceed 60°C., but up to 90°C. may be used without damage to the engine. The oil consumption should be between 8 and 16 pints per hour.

76. **Coolant system.**—A horseshoe type header tank, filled with 30% Glycol (D.T.D. 344A inhibited) and 70% distilled water, is mounted over the reduction gear of each engine. On the ground with engine running a small coolant discharge is normal, but not in flight.
    From B block on each inboard engine coolant is led to the cabin heating radiator, through which the flow of air is regulated by controls either side of the fuselage at the wireless operator's station.

77. **Hydraulic system**
    The accumulator, supplied by two pumps, has an air charging valve and a pressure gauge which should read 220 lb./sq.in. when there is no pressure in the system. Misleading pressure gauge readings will occur if the accumulator air pressure is incorrect. The gauge should read between 800–900 lb./sq.in. under working pressure, when the cut-out operates isolating the pumps and relieving them of unnecessary strain. The accumulator then provides the initial pressure to operate the various systems. When the pressure falls between 220–300 lb./sq.in. the pumps will automatically be cut in to operate the system and build up accumulator pressure again.

PART V—NOTES FOR FLIGHT ENGINEER

Hydraulic fluids.—The four fluids suitable are mineral oils. They may be mixed if additional supplies of that already in use are not available.

(i) Intava mixture 70% DTD585 + 30% DTD472.

(ii) Intava 695.

(iii) Mixture of two parts DTD44 and one part Intava 694.

(iv) DTD44.

Mixtures containing DTD44 will congeal at a higher temperature than the others. The first-mentioned mixture has the lowest congealing temperature.

Fuel jettison.—When the fuel jettison control is operated an air inlet valve is opened in the top of each No. 1 tank, and at the same time a valve is opened at the bottom of each tank, breaking a small retaining washer and releasing a spring-loaded stocking. Do not jettison below 100 gallons (*see* Part III, para. 60). The valves may be closed at any point but the stocking will not be retracted. When repacking, ensure that the stocking is dry and serviceable. Reseat the inlet and the outlet valves. Before fitting one of the spare shear washers carried in the housing, the spindle must pass centrally through the jettison valve.

78. **Pneumatic system**

The pressure (normally 300-320 lb./sq.in.) is controlled by a pressure regulating valve, which recommences charging when the pressure drops to 270-280 lb./sq.in. If the pressure drops to 130 lb./sq.in. or below, a pressure-maintaining valve closes, rendering the entire pneumatic system, with the exception of the brakes, out of operation. Therefore M supercharger ratio will be engaged, and the idle cut-off and radiator flaps will be inoperative. If pressure cannot be built up, to stop Merlin 28 or 38 engines, throttle down to minimum r.p.m. and close the master fuel cocks. This will drain the carburettor, which should be carefully primed to expel all the air from the fuel chambers before restarting. (*See* Para. 42 (vi).)

PART V—NOTES FOR FLIGHT ENGINEER

79. Electrical system

Four accumulators, connected in series parallel, giving a capacity of 80 ampere hours at 24 volts, are charged by two generators. Two ammeters on the main electrical control panel indicate the total generator output to services and battery charging. On the same panel a voltmeter indicates the state of the accumulators, reading 28–29 volts in flight under normal conditions and over 24 on the ground with engines stopped. On Lancasters I and III, spare fuses and a fuse location table are inside the hinged door. The main generator fuses are at the top inside the main control panel. Spare fuses are carried in the lid of the fuse box.

The switches below are for ground fault tracing and must be left on and not touched in flight.

On the front of the panel are two earth warning lights, a resin light switch (locked to prevent its movement) and the air bomber's station light switch of which the push button is for use on the ground in conjunction with the earth warning lights.

On Lancaster X, overload switches are fitted in the heavier circuits and flick off if the current becomes too great. Allow five or ten minutes for the service to cool down and turn the switch on.

80. Undercarriage failure

Should the red warning lights remain steady or the green lights not appear when the undercarriage is selected down, operate the warning light changeover switch. If this is ineffective select undercarriage up and return the selector down. If this is still ineffective check the accumulator pressure gauge reading. Should it be 300 lb./sq.in. there is probably no pressure in the system. Select the undercarriage up and down again. The pressure may drop between 200 and 300 lb./sq.in. then rise between 850 and 900 lb./sq.in. when the cut-out operates. If about 800 lb./sq.in. is indicated and the undercarriage appears to be down visually the indicator is probably faulty. If there is still some doubt use the emergency air

PART V—NOTES FOR FLIGHT ENGINEER

pressure to make certain the undercarriage is down, turning the cock off when the operation is complete. This conserves the air pressure for further use; if it is necessary to use the flaps emergency system the method of operation is described in para. 57.

81. **Gauges.**—The gauges on the engineer's panel in the Lancaster X are the four point type fitting into sockets and are therefore interchangeable.

AIR PUBLICATION 2062A, C & F—P.N.
Pilot's and Flight Engineer's Notes

PART VI
ILLUSTRATIONS AND LOCATION OF CONTROLS NOT ILLUSTRATED

LOCATION OF CONTROLS

Service	Location
Undercarriage warning horn test pushbutton	Behind pilot's seat.
Cross feed cock	On floor, just forward of the front spar.
Priming pump and cock (if fitted)	In each inboard nacelle.
Air intake heat control	Left of pilot's seat.
Radiator shutter switches	On starboard cockpit wall.
Supercharger gear change lever (on early Lancaster I)	To right of starboard master engine cocks.
Ground/flight switch.	On starboard side aft of front spar.
Cockpit heat controls	One each side of the fuselage just forward of the front spar. Two adjustable louvres in fuselage nose.
Oxygen master valve	At forward end of oxygen crate
Camera pushbutton control	On starboard rail of cockpit.
Reconnaissance flare stowage	On either side of fuselage forward of flare chute.
Flame floats or sea markers stowage	On either side of fuselage adjacent to flare chute.

FUSES

Location	Service
(a) Inside junction box at forward end of bomb aimer's compartment	Bomb gear fuses.
(b) Pilot's auxiliary fuse panel	Oil and radiator thermometer fuses.
(c) Navigation panel	Radio fuses.
(d) Mid turret position	Mid-upper and underturret, call lights, and, on early aircraft beam approach fuses.
(e) Main electrical control panel	General services.

KEY TO Fig. 1

INSTRUMENT PANEL

1. Instrument flying panel.
2. D.F. Indicator.
3. Landing light switches.
4. Undercarriage indicator switch.
5. D.R. compass repeater.
6. D.R. compass deviation card holder.
7. Ignition switches.
8. Boost gauges.
9. R.p.m. indicators.
10. Booster coil switch.
11. Slow-running cut-out switches.
12. I.F.F. detonator buttons.
13. I.F.F. switch.
14. Engine starter switches.
15. Bomb containers jettison button.
16. Bomb jettison control.
17. Vacuum change-over cock.
18. Oxygen regulator.
19. Feathering buttons.
20. Triple pressure gauge.
21. Signalling switchbox (identification lamps).
22. Fire-extinguisher pushbuttons.
23. Suction gauge.
24. Starboard master engine cocks.
25. Supercharger gear change control panel.
26. Flaps position indicator.
27. Flaps position indicator switch.
28. Throttle levers.
29. Propeller speed control levers.
30. Port master engine cocks.
31. Rudder pedal.
32. Boost control cut-out.
33. Signalling switchbox (recognition lights).
34. Identification lights colour selector switches.
35. D.R. compass switches.
36. Auto controls steering lever.
37. P.4 compass deviation card holder.
38. P.4 compass.
39. Undercarriage position indicator.
40. A.S.I. correction card hold.
41. Beam approach indicator.
42. Watch holder.

KEY TO *Fig. 2*
PORT SIDE OF COCKPIT

43. Bomb doors control.
44. Navigation lights switch.
45. D switch.
46. Auto controls main switch.
47. Pushbutton unit for T.R. 1196.
48. Seat raising lever.
49. Mixer box.
50. Beam approach control unit.
51. Oxygen connection.
52. Pilot's call light.
53. Auto controls attitude control.
54. Auto controls cock.
55. Auto controls clutch.
56. Brake lever.
57. Auto controls pressure gauge.
58. Pilot's mic/tel socket.
59. Windscreen de-icing pump.
60. Flaps selector.
61. Aileron trimming tab control.
62. Elevator trimming tab control.
63. Rudder trimming tab control.
64. Undercarriage control lever.
65. Undercarriage control safety bolt.
66. Portable oxygen stowage.
67. Harness release lever.

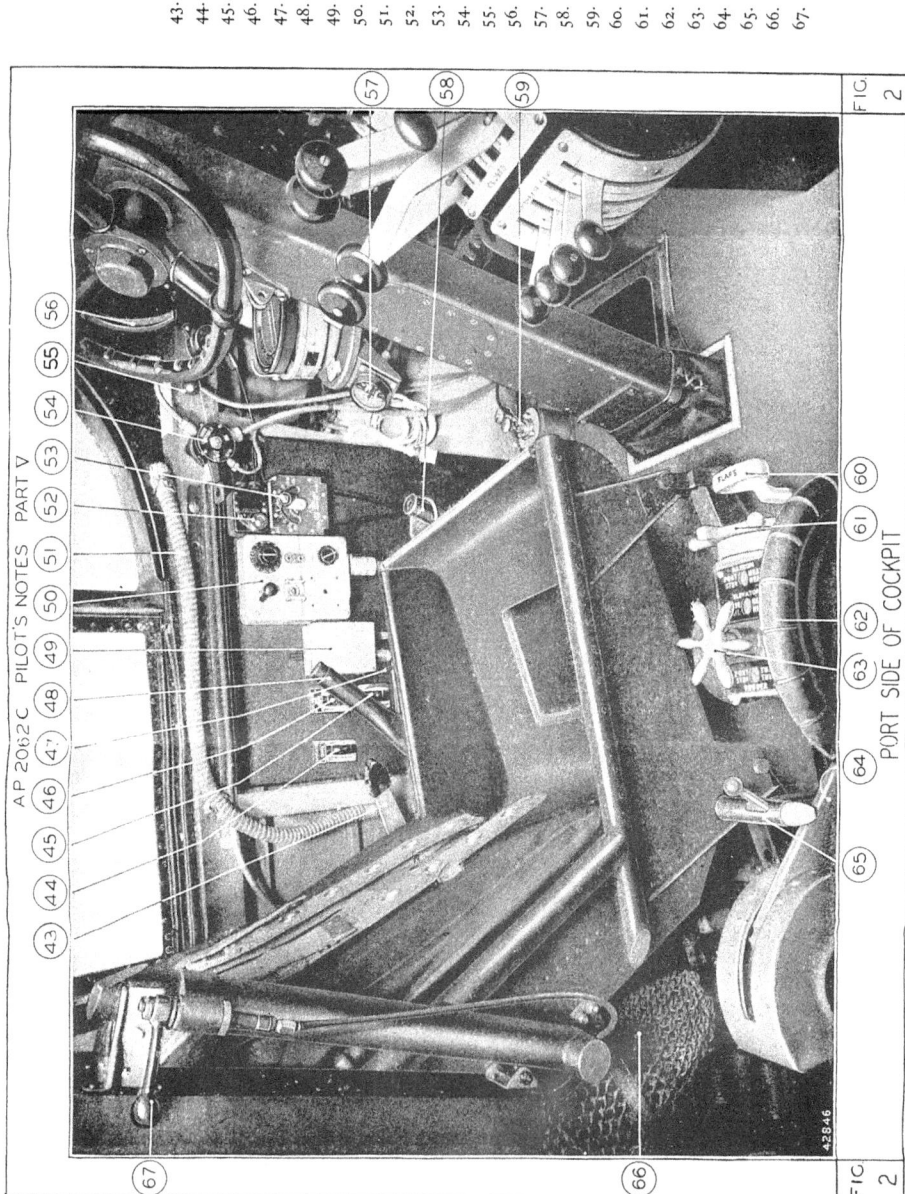

FIG. 2

FIG 3 — FLIGHT ENGINEER'S PANEL, LANCASTERS I AND III

KEY TO *Fig. 3*

FLIGHT ENGINEER'S PANEL

LANCASTERS I & III

68. Ammeter.
69. Oil pressure gauges.
70. Pressure-head heater switch
71. Oil temperature gauges.
72. Coolant temperature gauges
74. Fuel contents gauges.
75. Inspection lamp socket.
76. Fuel contents gauge switch.
77. Fuel tanks selector cocks.
78. Electric fuel booster pump switches.
79. Fuel pressure warning lights
80. Emergency air control.
81. Oil dilution buttons.

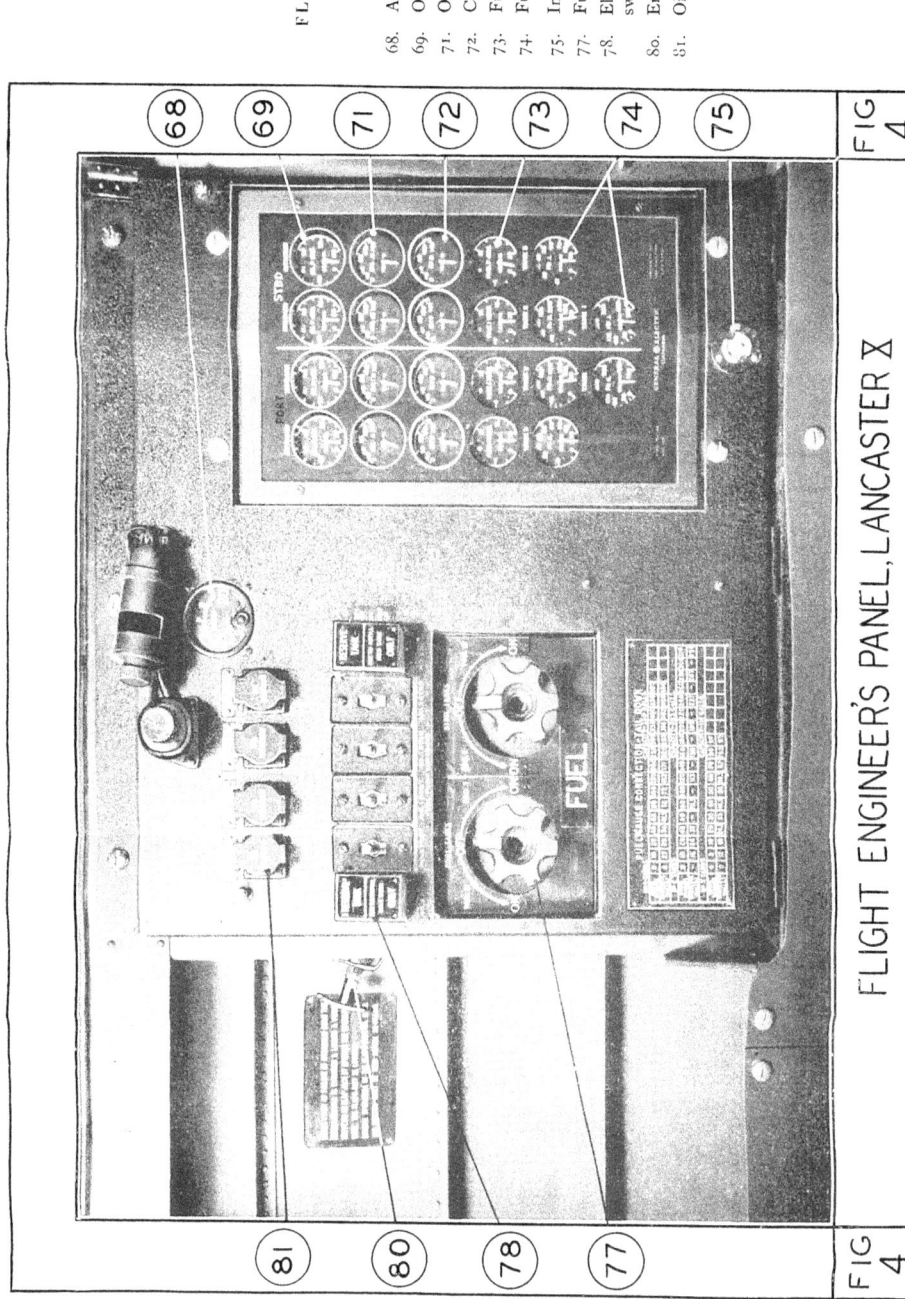

KEY TO *Fig. 4*
FLIGHT ENGINEER'S PANEL
LANCASTER X

68. Ammeter.
69. Oil pressure gauges.
71. Oil temperature gauges.
72. Coolant temperature gauges.
73. Fuel pressure gauges.
74. Fuel contents gauges.
75. Inspection lamp socket.
77. Fuel tanks selector cocks.
78. Electric fuel booster pump switches.
80. Emergency air control.
81. Oil dilution buttons.

FLIGHT ENGINEER'S PANEL, LANCASTER X

FIG 4

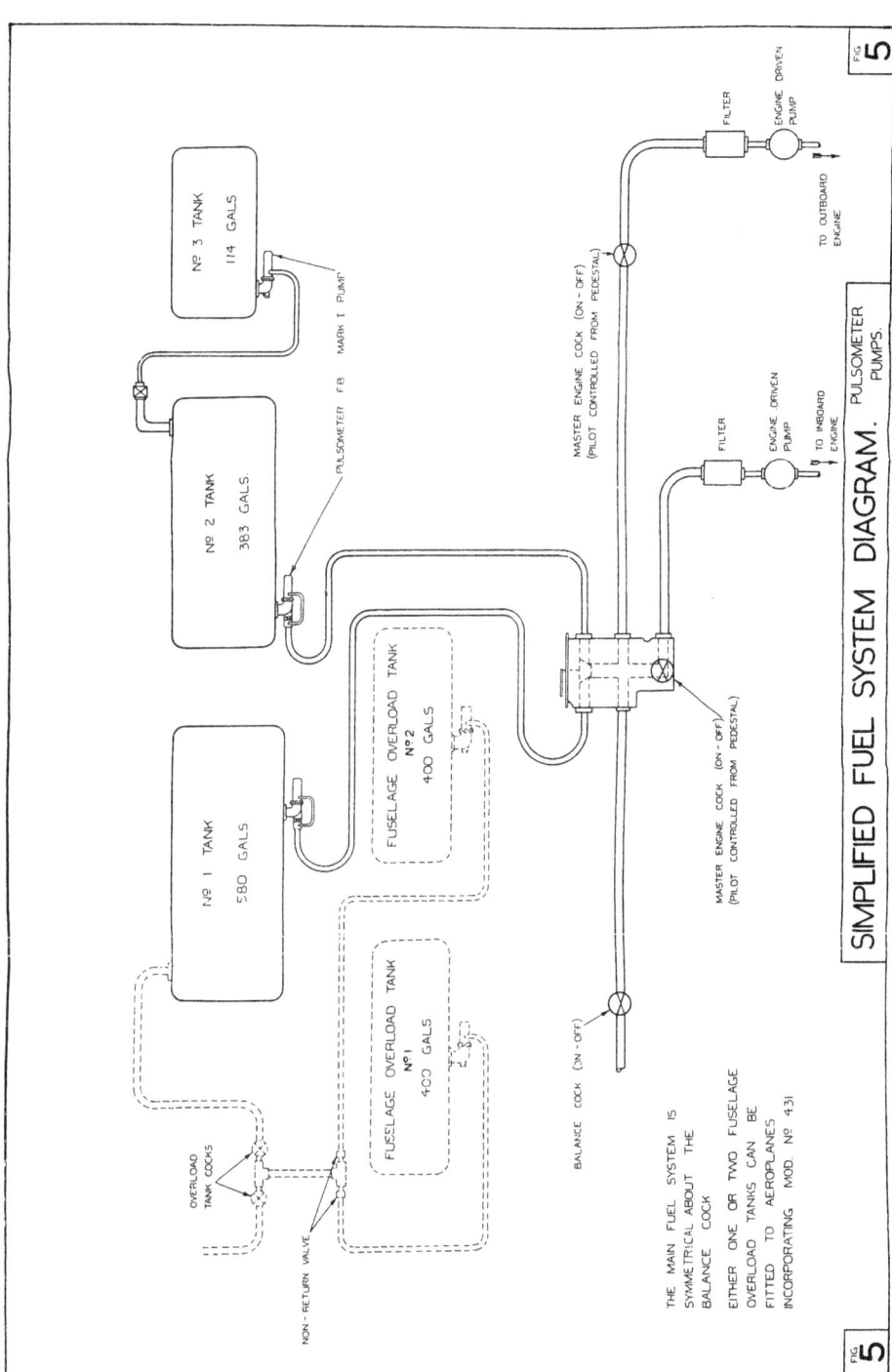

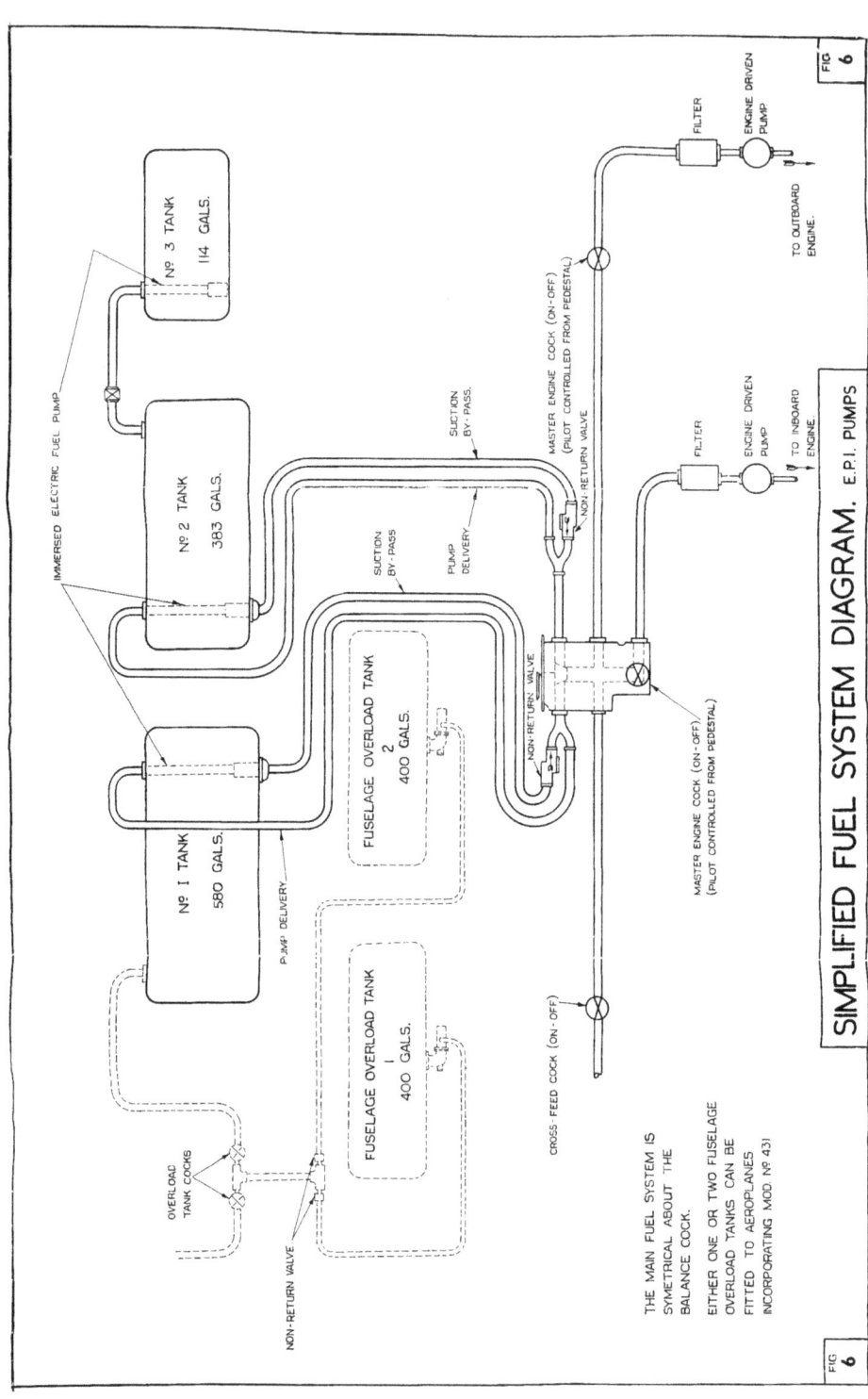

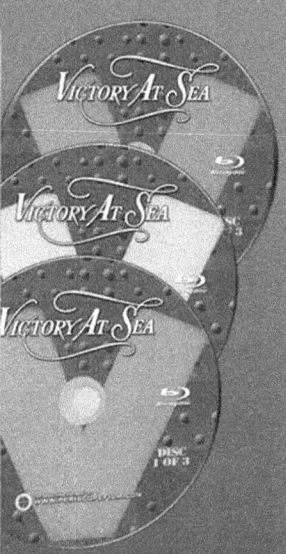

VICTORY AT SEA

IN HIGH DEFINITION
NOW AVAILABLE!

COMPLETE LINE OF WWII AIRCRAFT FLIGHT MANUALS

WWW.PERISCOPEFILM.COM

©2013 Periscope Film LLC
All Rights Reserved
ISBN#978-1-937684-52-5
www.PeriscopeFilm.com

www.ingramcontent.com/pod-product-compliance
Lightning Source LLC
Chambersburg PA
CBHW070656050426
42451CB00008B/374